Jaden Toussaint, The Greatest

The Greatest

Episode 2

THE LADEK INVASION

Plum Street Press

A Division of Yes, MAM Creations

Contents

Prologue

SPEAKS KINDNESS, OOZES CONFIDENCE.

JADEN
TOUSSAINT

Specializes in: knowing stuff. And also, ninja
dancing. He's really, really good at ninja dancing.

OWEN

Jokester and action expert.

Extreme dinosaur safari
bungee jumping?
Owen is your guy.

EVIE

Don't let the cuteness fool you.

This girl packs a punch.
Excels at: Being in
Charge.

Cicada hunter and math whiz.

SONJA

Also draws
excellent rainbows.

WINSTON

Can quote stats
from every World
Cup final.*

*that he has been alive for

Spots hurt feelings and distracted goalies from miles away.

Past president, International Society of Stealthy Felines. Resigned in scandal. Likes to be held like a baby and scratch stuff.

GRANDMASTER, CAT CHESS.

GRIS GRIS

ANIMAL OF MYSTERY

Guinea Pig never has the same name two weeks in a row.

This week, you can call him Betty.
He may or may not call you Al.

Baba:

Tall. Competitive. Competitive about being tall. Gives great piggy back rides. Prefers to be called "baba" which means "father" in Swahili. Does not speak Swahili.

Mama:

Loving. No nonsense. Most often seen reading fantasy books or experimenting with bean desserts. Gives good hugs.

Sissy:

Reader. Writer. Animal lover. Once gave up meat for 6 months, but was broken by the smell of turkey bacon. Plans to be the 1st PhD chemist to star in a Broadway musical.

Chapter 1
It's Just Spring

Jaden Toussaint
hated the spring.

Well, it's not exactly that he hated spring. It's more like he was on the fence about it. Spring was o-kay, but he just couldn't see why people thought it was so much better than winter.

Sure, he could eat snowballs in the spring.

And spring was the perfect time to fly his kite.

But there was no way any of the spring stuff could compete with his super-duper, ultra-deluxe winter coat.

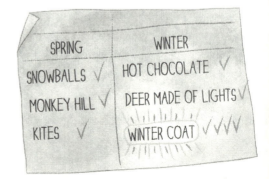

SPRING		WINTER	
SNOWBALLS	✓	HOT CHOCOLATE	✓
MONKEY HILL	✓	DEER MADE OF LIGHTS	✓
KITES	✓	WINTER COAT	✓✓✓✓

4 MONTHS EARLIER...

Jaden Toussaint's mom had let him pick it from a catalogue and it was pretty much the coolest coat ever. It was windproof, water resistant, and had a fancy, fuzzy lining that could zip in and out. It even had a hood.

SUPER-SWEET COAT

Creature Activation Device/ Head Warmer

Toranpu Card Pocket

Detachable Claws/Hand Warmers

The first time
JT put his new
coat on he had
been instantly
transformed
into a polar
bear—claws
and all. The
icy blasts
of winter
didn't stand a
chance against
his polar bear
armor. He
couldn't wait
to show his
friends.

WAIT UNTIL IT
GETS COLD, BABY.

5

There was just one problem:
the icy blasts of winter never came.

At first, he had waited patiently.

IS IT COLD ENOUGH TO WEAR MY COAT TODAY?

IS IT COLD ENOUGH TO WEAR MY COAT TODAY?

IS IT COLD ENOUGH TO WEAR MY COAT TODAY?

IS IT COLD ENOUGH TO WEAR MY COAT TODAY?

IS IT COLD ENOUGH TO WEAR MY COAT TODAY?

IS IT COLD ENOUGH TO WEAR MY COAT TODAY?

But after what seemed like *FOREVER*
(the grown-ups said it was only two
weeks, but it had to be longer than that),
Jaden Toussaint decided to take matters
into his own hands.

Science to the rescue!

All he had to do was learn everything there was to know about meteorology and he'd be able to fix the weather.

No problem, right?

Wrong.

By the end of the so-called "winter," Jaden Toussaint had learned an awful lot about meteorology, but he still hadn't discovered a way to make the weather do exactly what he wanted when he wanted it.

Despite his efforts, it had only gotten cold enough for Jaden Toussaint to wear his coat twice, and both of those times had been on the weekend. There's no school on the weekend! How was he supposed to magically transform into a polar bear for his friends when he never even got to wear his coat to school?

Then silly old spring arrived and JT's mother had packed his beautiful, extraordinary, magical coat into the back of his closet with all the other

winter things, never to be seen again. Possibly. Well, at least not until next winter and that was really, really far away.

Some people might think spring is puddle-wonderful, but Jaden Toussaint totally agreed with that poet:

It's Just Spring! Calm down, people. What's the big deal?

As far as he could tell, there was nothing about spring that could even sort of compete with a polar bear.

But then the mail came.

Chapter 2
DR. HOOOO?

There was one good thing about spring that Jaden Toussaint had forgotten: the spring issue of "Dr. Hoooo?"

"Dr. Hoooo?" was a comic book, but it wasn't just any comic book. It was the best, most action packed comic book in the world. It had a genius owl, a robot cat, a pygmy marmoset, tons of aliens from outer space, and a time machine tree that was bigger on the inside!

Dr. Hoooo? was so cool that just seeing it zapped all thoughts of JT's winter coat right out of his head.

Technically the comic book belonged to Sissy, and he had a vague memory of someone telling him something about it maybe being too scary for him, but when he saw it just lying there at the top of the pile he couldn't resist taking a quick peek.

At first it was nothing too terrible.

They did a little time travel...
Ran into a band of Chuckling Cherubs...
Dr. Hoooo? and Thumbs almost fell into an eternal fit of laughter...
Fe-Line stepped in to save them...
Dr. Hoooo? used his sonic hoot in a clever way to stop the Chuckling Cherubs for good...
Then Dr. Hoooo? and his two companions escaped and headed back to Time's Relative Elasticity Exploiter ^Time Machine^ (called *T.R.E.E.* ™, for short).

"What's everybody making such a fuss about?" JT thought to himself. "This comic book is not scary at all."

But when he turned the page he saw something so frightening that it made his hair stand even more on end.

Ladeks. Hundreds of Ladeks standing between Dr. Hoooo? and his time machine.

The Ladeks were Dr. Hoooo?'s most dangerous foes. They were the only ones who could get through the defenses of his time machine.

Jaden Toussaint had heard Sissy talking about the Ladeks before, but he had never imagined anything like this. They looked like worms, but they were covered with spiky armor and they could shoot laser beams from their scary little worm faces.

AHHHHHH!

FACE THAT
SHOOTS LASERS

POISON
SPIKES

LADEK

The next thing he knew, JT was shrieking in terror and Sissy was running into his room to check on him. She took one look at the Dr. Hoooo? comic book on the floor and immediately guessed what happened.

Sissy sat down next to Jaden Toussaint and put her arm around him. It felt good.

"Did you know that Dr. Hoooo? is a scientist just like you?" Sissy asked.

Jaden Toussaint nodded.

"So sometimes it takes him a few tries to solve his problem."

JT nodded again.

"Do you want me to read the rest of the comic book with you? It might make you feel better to see Dr. Hoooo? escape in the end."

Jaden Toussaint shook his head. He felt a little braver now that Sissy was in the room, but not *that* brave.

"Ok. I'll take it back to my room and put it on the top shelf where you can't see it. But if you change your mind..."

"No way," JT said. "I don't ever want to see one of those creepy Ladeks again."

She gave him one last hug, scooped up the comic book, and left the room.

After playing with Gris-Gris for a little while, Jaden Toussaint's thoughts turned to happier things like Toranpu cards and school. Before he knew it the Ladeks were completely out of his head. And good riddance.

KINDERGARTENERS ON CHESS BOARD

SISSY'S FRIEND (BOY)

SISSY

KID CHECKING FOR
GOODIES IN HIS LUNCHBOX

SISSY'S FRIEND (GIRL)

When Jaden Toussaint arrived at school the next morning, something seemed strange but he couldn't quite put his finger on it.

KID PICKING HER NOSE

Kids on the play equipment? Check.

Kids playing foursquare? Check.

PLAYING FOURSQUARE

Kindergarteners pretending to be chess pieces on the giant chess board? Check.

Then he saw what he had hoped he would never see again. Before he even realized it, he was screaming in terror.

WARNING

RED ALERT!

LADEK

WARNING

RED ALERT!

His shrieking brought Sissy over from where she had been talking to her friends.

"It's a La... a Ladek!" JT stammered. "And they're everywhere."

Sissy looked where Jaden Toussaint was pointing, but she didn't see any Ladeks. She saw something different.

"Toots," she said, "those aren't Ladeks. They're caterpillars. Buck moth caterpillars."

"I know," JT blurted out.

ARE YOU OKAY, TOOTS?

The thing is, he didn't really know. I mean, now that Sissy said they were caterpillars they did kind of look like caterpillars, and he felt silly insisting they were aliens who had traveled from another dimension, so he just went along with it.

"Ok, then. Do you want to take a closer look with me?" Sissy asked.

Jaden Toussaint was a little nervous. He wanted to say yes. After all, every good scientist knows about the power of observation. Having a closer look couldn't hurt. But even though he knew that, and even though he believed Sissy when she said the strange little creatures were just caterpillars, he still felt wary.

Seeing JT was a little hesitant, Sissy crouched down first. She was less than a foot away from the thing, and seemed to be watching it intently.

Wait a minute. Wait just one minute. Was Sissy out-science-ing him? No way! No way was Jaden Toussaint going to let that happen. He walked over to where Sissy was and knelt down beside her.

Up close the buck moth caterpillar didn't look nearly so scary. It stretched and rippled as it moved, kind of like an accordion on legs. Sure, it was covered in tiny little spikes, but the spikes were actually kind of cute. He wondered if they just <u>looked</u> hard or if they actually <u>were</u> hard.

JT reached out to touch it, but before he could Sissy had pulled his hand away.

"First of all," she said in her grown-up voice, "when you're doing an investigation you never touch unfamiliar creatures with your bare hands."

"I know," JT said.

Truthfully, he had forgotten about that safety rule, but he was so frustrated with himself for forgetting it, that "I know" was the first thing that came out.

"Second of all," Sissy continued, "I only know two things about this kind of caterpillar. 1. They come out every spring. 2. They sting."

They STING?!? What?

Jaden Toussaint was just about to ask Sissy so many questions like:

Why'd you let me put my face so close? and
Seriously? Every spring? and
How sure are you that it isn't a Ladek?

Then the bell rang and everyone ran to line up for the start of the school day.

Chapter 4

BOOTS

JT, Evie, Sonja, Winston, and Owen were all assigned to the science center that morning. Normally, Jaden Toussaint was the kind of guy who didn't like messing around before his work was done, but he thought that since he was trying to prevent the spread of a possible alien invasion Miss Bates might forgive him for being off-task just this once.

He told them everything. The comic book. The Ladeks. How the Ladeks bore an uncanny resemblance to what Sissy said were buck moth caterpillars. And how, even though he thought Sissy was probably right about them being caterpillars, he didn't want to rule out the possibility that they were evil alien invaders... you know... just in case. And, as proof of their possible alien nature, he told his friends about the stinging.

"Lots of creatures sting. That doesn't mean they're aliens," said Sonja.

"I'm pretty sure they're just caterpillars," Owen said. "My dad and I have seen them before."

"Couldn't we just talk to it?" asked Winston.

"Guys, I hate to be the one to say this, but whether it's an alien or a caterpillar, there's really only one thing we can do," said Evie. "Squish it."

JT thought for a moment. Maybe squishing it wasn't such a bad idea. Sometimes in games like Toranpu or chess you had to attack first to stop your opponent from attacking you. Wasn't this pretty much the same thing?

Besides, Evie was tough. If anybody could stomp a caterpillar it would definitely be Evie.

"But the stingers," JT said finally. "How can you squish it without getting stung?"

"Because I have these," Evie said, pointing at her boots.

THESE BOOTS WERE MADE FOR STOMPING.

Recess finally came and Evie learned something new:
Stomping on twigs and dead leaves was one thing, but stomping on a caterpillar was quite another.

Evie took one look at the caterpillar and had a change of heart. "He's so cuuute!" she exclaimed. "Look at him with his spiky little armor. He's adorable!"

"His spiky, stinging armor," Sonja corrected. "And it looks like he's got some friends."

JT and his friends looked around. There had only been a few caterpillars on

the yard this morning, but now there were at least 10. Maybe more.

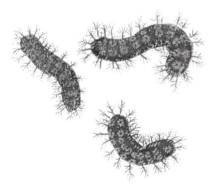

Suddenly, Owen came running up. As the group's best spy, Owen had gone on a mission to gather information. He was running toward them so fast that they all knew it must be serious.

"Guys!" Owen panted, trying to catch his breath. "I overheard [pant] the teachers [pant] talking [pant pant]..."

"And...?" They were practically shouting at Owen, but the suspense was killing them. There was a faint cry from the other side of the yard, but the all of them were concentrating on Owen.

"They've called out...
THE EXTERMINATOR."

DUNH-DUNH-
-DUUUUNH!

Chapter 5
EXTERMINATE! EXTERMINATE!

Recess ended abruptly and all the classes lined up quickly and went inside.

Miss Bates gathered the whole class on the rug for circle time, which was weird because it wasn't circle time.

"I have some bad news," Miss Bates said. "Kae, one of our friends from another class, got stung by a caterpillar at recess today."

Everyone had questions. Miss Bates did her best to answer them all.

When everyone was quiet again, Miss Bates said: "There are two more things you need to know. The first thing is that the school has called an exterminator to come get rid of the caterpillars. The exterminator will be here on Saturday."

Exterminate? That seemed a bit harsh. Just because they look like Ladeks doesn't mean you have to exterminate them.

Miss Bates continued. "The second thing is that in the meantime we're going to have indoor recess." Everybody groaned, including Jaden Toussaint. That was the last straw!

WRONG STRAW, DUDE.

Miss Bates thought that they were groaning because they didn't like indoor recess, so she started talking about the fun things they could do inside and how important it was to keep everybody safe.

Jaden Toussaint actually liked having indoor recess (it gave him an extra chance to use the computer station), so that's not what he was groaning about. He was still thinking about The Exterminator.

The Exterminator couldn't be the only way to stop the caterpillars. There weren't even that many of them.

What would Dr. Hoooo? do? JT hadn't finished the comic book, but he was pretty sure that Dr. Hoooo? would think like a scientist. And Jaden Toussaint knew all about thinking like a scientist.

All he had to do was:

-Figure out a way to save the caterpillars

-Stop everyone from getting stung

-Make it safe to play on the playground

-And work it all out before Saturday

Whew. Now that he thought about it, it did seem like a lot. He still thought that he could do it, but he knew he couldn't do it alone. Luckily, he didn't have to.

Chapter 6
FIVE HEADS ARE BETTER THAN ONE

Winston, Sonja, Owen, and Evie convinced their parents to let them hang out at JT's house after school. The grown-ups kept calling it a playdate,

but they all knew the truth. They had work to do!

Next Stop: Research.

Mama made popcorn. Sissy helped them print information from the internet. Jaden Toussaint pulled out every one of his nature books and magazines.Then he and his friends got down to business.

Two hours later they knew a whole lot more about buck moth caterpillars.

Winston found out that some caterpillars communicate with their rear ends.

Evie learned that the caterpillar's cute little spikes were really hollow spines attached to a venom gland. Awesome.

Owen found a cool chart of the lifecycle of a buck moth.

Jaden Toussaint read that buck moth caterpillars only crawl to the ground to find a place to change into adults.

Sonja learned that sometimes the wind blows the caterpillars off course.

Lots of information. Zero ideas.

Thanks to his gigantic brain, most ideas just popped right into JT's head without him trying. But this caterpillar thing was a toughie, and they were running out of time. It was almost time for everyone to go home.

For stubborn problems like this one, there was one surefire way to kick their brains into top gear.

3 MINUTE

WARRIOR WORM

HOOTING OWL

SYNCHRO

RAINBOW

DANCE PARTY!

NINJAS ON TIPTOE

SHADOW

SNEAKY LITTLE ELF   LOOK OUT!

I think it's...
Yes!

Then, out of nowhere, it started. That swirly, whirly, zinging feeling he got whenever he was on the verge of a brilliant idea. And just like that, Jaden Toussaint knew what to do.

Chapter 7
O.S.C.A.R.

The plan was set. Operation Save Caterpillars and Recess was in motion. Everybody had a part to play.

Evie was in charge of warning signs.

Owen's job was to protect the caterpillar paths.

Sonja was in charge of leaf blowers.

Winston was in charge of teaching everyone about caterpillar safety.

LIVING WITH CATERPILLARS

A GUIDE TO COEXISTING

WITH BUCK MOTH CATERPILLARS

BY WINSTON P., PHD*

DIRECTOR OF EDUCATIONAL OUTREACH

*PEACEFUL, HELPFUL DUDE

Jaden Toussaint was in charge of body armor.

You see, one thing JT and his friends had learned about buck moth caterpillars is that they don't use their painful spines on purpose. The spines are just an automatic layer of protection to keep them safe from predators.

Well, if the caterpillars had an automatic layer of protection, then humans should, too, right? Jaden Toussaint knew just the thing.

BODY ARMOR

He and mama sent a message to
everyone asking them to help out with
Operation Save Caterpillars and Recess
by wearing their coats to school. JT
thought the message should say "polar
bear armor" instead of "coat" (coat
sounded so plain), but Mama said that
might be confusing.

By the next morning Jaden Toussaint
was so excited that he woke Mama up
a teensy bit early to see if everyone had
gotten the message.

It looked like they had.

And best of all, Jaden Toussaint finally, FINALLY got to be a polar bear at school.

It turned out that he wasn't the only one who had been looking forward to wearing his coat.

47

Even though it was kind of hot outside, wearing their coats and saving caterpillars seemed to make the whole school happy. JT could tell they were happy because they were all smiling.

The kids wearing coats were smiling.

The parent volunteers were smiling while they blew caterpillars into the safe zone with leaf blowers.

Even the warning signs seemed to smile as they sparkled in the morning sun.

With everyone pitching in to help, the playground practically looked like a party.

Do you know what? That did the trick.

When the principal, Mrs. Nelson, saw how hard everyone was working to save the caterpillars (and recess), she was so inspired that she cancelled the exterminator and decided to come up with a new plan.

Miss Bates' class was there to help.

After a little trial and error, they came up with a plan to protect the whole playground at once: a giant tarp!

If they stretched it really tight and tilted it toward the ground at one end, the buck moth caterpillars that got blown out of the oak trees would slide away from the playground instead of landing on it.

Mrs. Nelson was so proud of Jaden Toussaint and his friends that they each got a sticker.

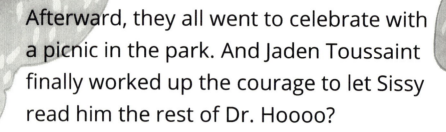

Afterward, they all went to celebrate with a picnic in the park. And Jaden Toussaint finally worked up the courage to let Sissy read him the rest of Dr. Hoooo?

And guess what? Even with all his fancy time traveler technology, Dr. Hoooo? had come up with the pretty much the same solution as JT and his friends.

Who needed time travel? With friends like these, he was already the greatest.

53

Epilogue

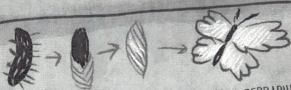

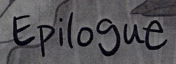

-MAMA SAYS I CAN'T BUILD A BUCK MOTH TERRARIUM. OAK TREES TOO BIG. MAYBE MINIATURIZE THEM?

-**IN** JUST SPRING = Thumbs Fine

-ADULT BUCK MOTHS HAVE NO MOUTHS!! SAD. THEY WILL NEVER TASTE HOT CHOCOLATE.OR SNOWBALLS.

BUCK MOTH CATERPILLARS		VS. LADEKS
HABITAT	OAK FORESTS VERNAL,	THE PLANET OF ETERNAL SPRING
FOOD	MAINLY OAK LEAVES	T.R.E.E. ™ S
PRIMARY DEFENSE	SPINES ATTACHED TO VENOM SACKS. ONLY HURTS IF TOUCHED.	LASER FACES
STATUS	FACT	(PROBABLY) FICTION

About the Illustrator

Marie Muravski was born in a city deep in Siberia. There were no bears on the streets of her city, but it was surrounded by the most beautiful pine forests. Those pine forests inspired her to draw as a little girl and she never stopped. Now that she is all grown up, being surrounded by nature still inspires her art.

She and her husband travel the world looking for new beautiful places to inspire them, making art, and just being awesome together.

You can find her at:

www.facebook.com/MarieMuravski

THIS BOOK? ....

About the Author

Marti Dumas is a mama who spends most of her time doing mama things. You know - feeding ducks in parks, constructing Halloween costumes, facilitating heated negotiations, reading aloud, throwing raw vegetables on a plate and calling it dinner, and shouting, "watch out!" whenever there are dog piles on the walk to school.

Sometimes she writes, but only very occasionally and in the early morning. And, yes. She really does really, really like fantasy books. A lot.

You can find her at:

www.MartiDumasBooks.com

JADEN TOUSSAINT, THE GREATEST

EPISODE 1: THE QUEST FOR SCREEN TIME

Written by Marti Dumas

Illustrated by Marie Muravs

AVAILABLE NOW
Ever Dream of Being a Wolf?
JALA
and the
WOLVES
by Marti Dumas

For printables, ninja dance music, and more, visit:

www.MartiDumasBooks.com

Authors love reviews.
We eat them up like pizza
for breakfast.

Yum!

CPSIA information can be obtained
at www.ICGtesting.com
Printed in the USA
LVOW10*1451020317

525945LV00007B/127/P